1%

D0475166

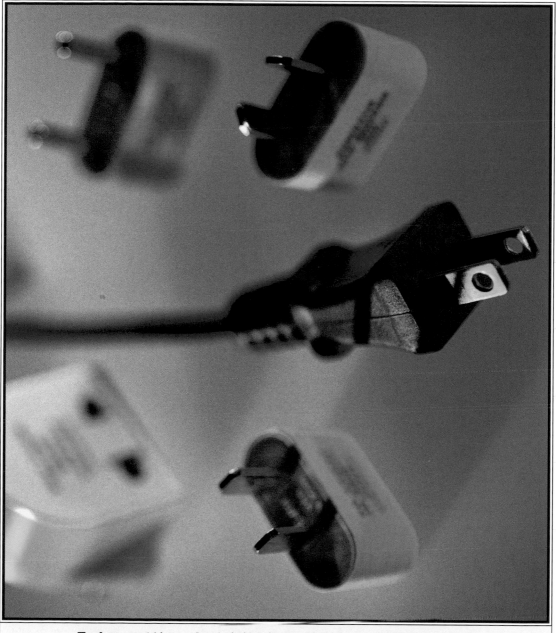

Today, getting electricity is as simple as "plugging in"

Electricity

Jean Allen

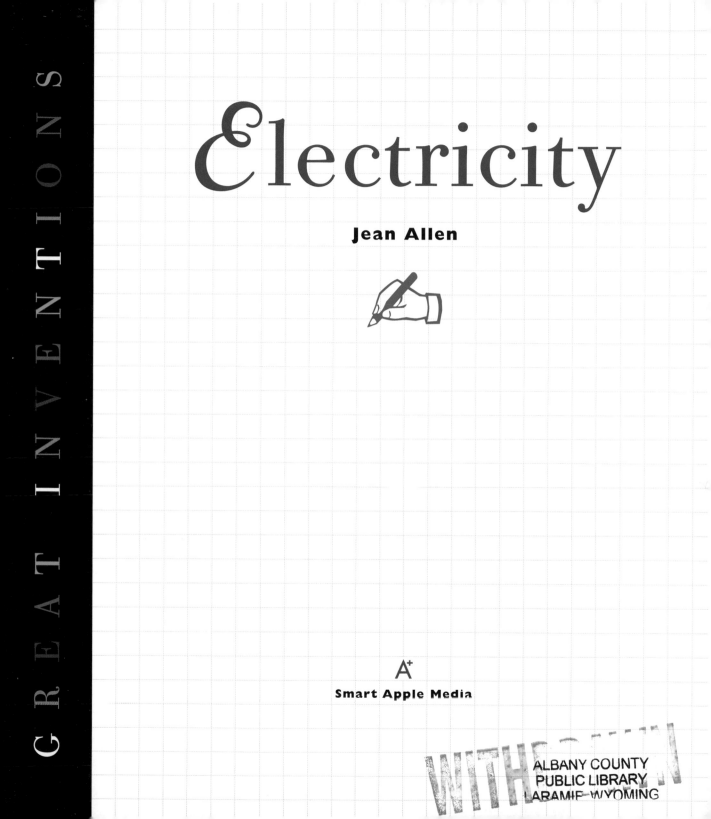

A⁺

Smart Apple Media

COPYRIGHT

✍ Published by Smart Apple Media

1980 Lookout Drive, North Mankato, MN 56003

Designed by Rita Marshall

Copyright © 2004 Smart Apple Media. International copyright reserved in
all countries. No part of this book may be reproduced in any form without
written permission from the publisher.

Printed in the United States of America

✍ Photographs by Corbis (Bettmann, Digital Art, Philadelphia Museum of Art),
Richard Cummins, JLM Visuals (Doug Reid), Tom Myers, Tom Stack & Associates
(David Young), Unicorn Stock Photos (Novastock)

✍ Library of Congress Cataloging-in-Publication Data

Allen, Jean, 1964– Electricity / by Jean Allen.

p. cm. – (Great inventions) Includes bibliographical references.

Summary: Briefly describes what electricity is, how various people developed
a better understanding of this form of energy, and how electricity is generated.
Includes a related activity.

✍ ISBN 1-58340-320-5

1. Electric engineering–Juvenile literature. 2. Electricity–Juvenile literature.
[1. Electricity.] I. Title. II. Great inventions (Mankato, Minn.).

TK148.A44 2003   621.3–dc21   2002042789

✍ First Edition 9 8 7 6 5 4 3 2 1

Electricity

What Is Electricity?

A lightning bolt rips apart the night sky. A baseball game is played beneath blazing stadium lights. A woman uses a flashlight to look for a lost button beneath a couch. What do these scenes have in common? They are all examples of a mysterious form of energy called electricity. ✎ Electricity is part of nature. It is created within atoms—tiny, tiny particles that make up all matter. Atoms have three parts. Protons and neutrons join together to form the center, or nucleus, of the

Lightning is a natural and dangerous form of electricity

atom. Electrons orbit (circle) around the nucleus. An atom is a little like our solar system. The nucleus is like the Sun, and the electrons are like the planets that orbit around the Sun.

The protons and neutrons of an atom usually stick together. But electrons can be gained or lost. This changes the **charge** of the atom. An atom with more electrons than protons has a negative charge. An atom with more protons than electrons has a positive charge. Electricity

Lightning is caused by ice crystals high in the clouds knocking against each other, creating an electrical charge.

A diagram of an atom, with Earth as the nucleus

Electricity lets human activity continue even at night

is created by the movement of electrons among atoms.

Discovering Electricity

Around 600 B.C., a Greek scientist known as Thales of Miletus made a fascinating discovery. He rubbed a piece of amber (the hardened sap of pine trees) against a piece of fur. Feathers stuck to the amber as if by magic. Thales did not know it, but he was observing **static electricity**. Little else was learned about electricity until 1600, when Englishman William Gilbert performed experiments similar to Thales's and wrote a book about them. Electricity became a very important

area of study. One of the best-known experiments was

conducted in 1752 by American inventor Benjamin Franklin.

He tied a key to a kite and flew the kite during a thunderstorm.

Inventor Benjamin Franklin attaching a key to his kite

When the key became charged with electricity, Franklin had proof that lightning is a flow of electricity in nature.

Another important step in understanding electricity was taken in 1800, when an Italian scientist named Allesandro Volta invented the first electric **battery**. This opened the door to other discoveries, because scientists now had an **The story _Frankenstein_ (1818) is about a scientist who uses electricity to bring a dead body to life.** easier way of storing and working with electricity. By the end of the 1800s, many electrical devices had been invented, including the electric motor, telephone, and light bulb.

Electricity at Work

The more scientists learned about electricity, the more they were able to harness the energy and put it to use.

An operator working an old telephone switchboard

Beginning in the early 1900s, electricity was created in power

plants and delivered to homes through wires. Life changed in

many ways. Electric lights replaced kerosene lamps.

Refrigerators kept food cold. Water **Benjamin Franklin's**
experiments with

heaters let people take warm baths **lightning were**
risky. A Russian

without having to heat water over a **scientist was killed**
trying a similar

fire. Today, our homes are filled **experiment.**

with electrical devices such as computers, radios, air

conditioners, televisions, and microwave ovens. Everything

Electricity made the invention of light bulbs possible

that uses a battery—such as a toy, cell phone, or watch—is using

electricity. Even items that do not use electricity were probably

made with machines or tools that use electricity.

Making Electricity

In order to make electricity, other sources of energy are

needed. Today, most of the world's electric power comes from

coal, oil, **nuclear** energy, and natural gas. These energy

sources are burned, and the heat boils water to make steam.

The steam moves a spinning blade called a turbine, which is

connected to a generator—a machine that makes electricity by

turning a **magnet** inside a coil of wire. ✐ No method of

making electricity is perfect. Coal, oil, natural gas, and nuclear

fuel create pollution. This is especially troubling because our

Electricity is produced in power plants like this one

energy needs are rising every year. Better sources of energy are

wind, water, and solar (sun) power. These do not harm the

environment as much, but they are more expensive and are not

available in all areas. Producing **Alexander Graham Bell invented the telephone in 1876. Just three years later, Thomas Edison invented the light bulb.**

electricity in a responsible way will be

a huge challenge in the years to come.

It is a challenge we must meet. From

desk lamps to stadium lights, from calculators to space shuttles,

electricity is an essential part of our lives.

Space shuttles need electricity to fly into outer space

Charging Up

Try this experiment to create static electricity and see how positive and negative charges affect each other.

What You Need

A balloon
A wool sweater

What You Do

1. Blow up the balloon and tie it shut.
2. Rub the balloon on the wool sweater. (You can rub it on your hair if you do not have a sweater.)
3. Hold it up to a wall. The balloon will stick to the wall.

What You See

When you rub the balloon, it picks up extra electrons from your sweater or hair. This gives it a negative charge. The balloon sticks to the wall because the wall is positively charged. You can also try picking up feathers, bits of paper, or other light objects like Thales of Miletus did with his amber. Or rub two balloons on the sweater and try holding them near each other. What happens? Why?

Static electricity can make your hair stand straight up

Index

Words to Know

battery (BAT-uh-ree)—a device that stores chemicals that can be turned into electrical energy

charge (CHARJ)—electrical energy that is either positive or negative; positive charges are attracted to negative ones, and vice versa

magnet (MAG-nit)—an object that attracts, or pulls toward itself, anything made of iron or steel

nuclear (NOO-klee-ur)—energy created by splitting certain kinds of atoms

static electricity (STAT-ik ee-lek-TRIH-sih-tee)—electricity that collects in an object and stays there

Read More

Berger, Melvin. *Switch On, Switch Off*. New York: HarperTrophy, 2001.

Glover, David. *Batteries, Bulbs, and Wires*. New York: Houghton Mifflin, 2002.

Tocci, Salvatore. *Experiments with Electricity*. New York: Children's Press, 2001.

Internet Sites

Science Learning Network
http://www.miamisci.org/af/sln/

Science Made Simple
http://www.sciencemadesimple.com/what.html

Energy Quest, California Energy Commission
http://www.energyquest.ca.gov/index.html